Maximum weights in load lifting and carrying

OCCUPATIONAL SAFETY
AND HEALTH SERIES
No. 59

MAXIMUM WEIGHTS IN LOAD LIFTING AND CARRYING

INTERNATIONAL LABOUR OFFICE   GENEVA

ISBN 92-2-106271-6
ISSN 0078-3129

First published 1988

Printed by the International Labour Office, Geneva, Switzerland

CONTENTS

I. Introduction

The purpose of this document is to present a summary of the legislation and practice concerning the limitations of weight in manual lifting and carrying of loads adopted in various ILO member States.

In recent times, a number of publications, including some outstanding investigations, have been issued on physiology, ergonomics, hazards and recommended techniques of manual lifting, carrying, pulling and pushing loads. These data are easily accessible, particularly by means of information systems such as the International Occupational Safety and Health Information Centre (CIS) of the ILO. However, data on statutory provisions and standards adopted in member States are not easily available although they may be of primary interest to government departments, employers' and workers' organisations.

The sources of the present compilation were:

(a) the replies to the ILO investigation on special protective measures for women workers in the field of working conditions collected by the ILO regional offices in 1985;

(b) the papers submitted to the International Symposium on Ergonomics in Developing Countries (Jakarta, 18-21 November 1985);

(c) the replies to the Call for Information on ways to reduce back injuries at the workplace in the framework of the International Occupational Safety and Health Hazard Alert System (1987);

(d) laws and regulations available to the ILO.

The document cannot therefore pretend to be fully exhaustive at this stage. Further data, which will be notified to the Office, will be added to the compilation in order to keep it up to date and more informative.

The material has been classified as follows:

- general provisions;

- statutory limits of loads to be lifted and carried by adult male workers;

- statutory limits of loads to be lifted and carried by adult women workers;

- statutory limits of loads to be lifted and carried by young workers and children;

- no limiting provisions;

- recommendations for and information on practice.

II. General provisions

1. <u>Several member States have issued general provisions concerning health protection in manual handling of loads, applying to all workers</u>:

- <u>Austria</u>: "Workers may be allocated to lifting, carrying and transport of loads only in accordance with their constitution and physical strength" (1983).

- <u>Canada</u>: Material Handling Regulations under the Labour Code state: "Where, because of the weight, size, shape, toxicity or other characteristics of a material or object, the manual handling of that material or object may endanger the safety or health of an employee, the employer of any such employee shall ensure to the extent that is reasonably practicable, that the material or object is not handled manually."

In Quebec, Canada at the level of the provincial jurisdiction, the Safety Code for the Construction Industry states: "Mechanical apparatus shall be provided and used for carrying material when the safety of the worker is jeopardised."

- <u>Guatemala</u>: General Regulations on Occupational Hygiene and Safety provide that loads transported by workers shall be suited to their physical powers, regard being had to their character, weight and volume of the load and the distance and road to be travelled.

- <u>Ireland</u>: the Factory Act provides that no person shall be required to lift, carry or move a load so heavy that it may cause him injury.

- <u>New Zealand</u>: the Factories and Commercial Premises Act, 1981, section 26 on Carrying of Heavy Loads, states: "No worker in an undertaking shall be required to lift or carry a load which is so heavy that it could injure him (her)."

- <u>Pakistan</u>: according to the Factories Act: "No person shall be employed in any factory to lift, carry or move any load so heavy as to be likely to cause him injury."

- <u>Sri Lanka</u>: the Factories Ordinance prescribes that no person shall be employed to lift, carry or move any load as heavy as to be likely to cause injury to him.

- the United States (Oregon): the Wage and Hour Commission of Oregon prohibits requiring any employee to lift excessive weights.

2. **In a few member States, general provisions apply to the manual transport of loads by adult women workers:**

- Egypt: a Ministerial Decree in application of article 153 of the Labour Law prohibits to women the work of loading and unloading of goods at docks, wharfs, ports and warehouses.

- Malaysia: "No expectant mother shall be required to carry, push, or pull any load."

- Netherlands: the Labour Decree states: "A woman may not carry out any work which involves lifting, pulling, pushing, carrying or moving a load in any other way if this work apparently or in the opinion of the Head of the Factories Inspectorate: (a) requires too great an effort, (b) is a danger to her health for any other reason."

3. **In some member States, general provisions apply specifically to young working persons and children:**

- Austria: "Young workers may not be employed at lifting, carrying, pushing or moving - with or without technical means - loads involving an intolerable demand on the organism."

- Egypt: under 17 years of age, the work of loading and unloading of goods at docks, wharfs, ports and warehouses is forbidden.

- Luxembourg: the law concerning the protection of children and young workers prohibits any work of young people requiring a physical effort exceeding the worker's powers.

- Netherlands: the Labour (Youth) Decree states: "A young person may not carry out any work which involves frequent lifting or carrying of heavy loads or which means adopting an inconvenient posture continuously for a longer period of time", and "A young person may not carry out any work, which, because of the effort required, would result in a workload which could be regarded as being unacceptable for him."

4. **Some legislation contains special provisions according to the sex of the young working person:**

- Hungary: boys under 14 and girls under 16 years of age may not be employed under any circumstance in loading and transport of loads and boys between 14 and 16 and girls between 16 and 18 years of age may not be employed in permanent transport of loads. Furthermore, boys between 14 and 16 and girls between 16 and 18 years of age may not carry unaided any load over slopes exceeding 2 per cent.

5. A few member States have adopted provisions aiming at the health protection of special categories of workers against the hazards involved in the use of certain equipment in load carrying and/or lifting, such as:

- Cameroon: girls and young women under 18 years of age may not carry loads by means of hand barrow and two-wheeled carts.

- Hungary: young workers between 16 and 18 years of age may not be employed for mine-car transport and may not regularly transport materials by means of hand barrow.

III. Limitations of weight

1. Adult male workers

In the following pages, member States limiting the maximum weight to be manually lifted and carried by one adult male worker are listed at the beginning with the highest limits and going on with States that have adopted lower limits.

National legislation on maximum weights

(a) maximum weight for adult males

Country	Conditions	Maximum weight
Czechoslovakia	Good gripping facilities for both hands	50 kg
Ecuador	None	60 kg
Greece	For carrying meat of slaughtered animals	100 kg
Mozambique	None	55 kg
Pakistan (Punjab)	None	90 kg
Philippines	In construction work (continuous lifting)	50 kg

(b) Some national laws contain more detailed maximum weight provisions, such as:

Brazil	For carrying, for a distance of 60 metres	60 kg
	For lifting by a worker on his own	40 kg

- Colombia: The maximum weight that may be carried on
  the shoulders is .... 50 kg
  The maximum weight that may be lifted is .... 25 kg

- German Democratic Republic: Lifting regulations for
  adult male workers take into account both the workload and the time spent on lifting as shown below:

Frequency of lifting and carrying per shift in % of shift time	Heavy to very heavy work	Moderately heavy work	Light to moderately heavy work
Less than 8 (sporadic)	45 kg	38 kg	33 kg
8-30 (frequent)	25 kg	18 kg	12 kg
More than 30 (constant)	15 kg	12 kg	8 kg

- Federal Republic of Germany: A recommendation on limits for
  lifting and carrying loads has been transmitted for
  application by letter of 1 October 1981 by the Federal
  Ministry of Labour and Social Affairs to the Safety and
  Health Inspection of the Federal State Governments (Länder),
  and to the Confederation of German Employers' Association
  (BDA) and the German Confederation of Trade Unions (DGB)
  (Bundesarbeitsblatt 11/1981, s. 96). The recommendation
  refers to a study of Professor T. Hettinger, which was
  published in 1981 by the Federal Ministry of Labour and
  Social Affairs. The study provides a table containing
  limits for lifting and carrying loads.

8   6875d/v.5

Table 1: Limits for lifting and carrying loads.
Lifting strengths for women and men

Age	Admissible load/kg Frequency of lifting and carrying			
	Occasionally		More frequently	
	Women[1]	Men[1,2]	Women[2]	Men[2]
15-18 years	15	35	10	20
19-45	15	55	10	30
Over 45 years	15	45	10	25

[1] Unshaded = limits which must normally not be exceeded without health risk.

[2] Shaded = values recommended from an ergonomic point of view.

This table shows the admissible weight limits for three age classes. These were further subdivided by the frequency of lifting and carrying, but here again only in broadly indicative ranges, i.e. "occasionally" (less than twice per hour) and "more frequently" (more than two to three times per hour).

Within the range covered by the word "occasionally", the limitation of admissible weight is supposed to prevent damage to reproductive organs; for young males, the limit is determined by the tolerance of the spinal column and the objectively determinable individual strength.

Within the range "more frequently", the limiting factor for women is, however, no longer the status of the reproductive organs, but the determinable tolerance of the muscular and the cardio-pulmonary systems.

The question of what is a damaging overload can in this case be answered in terms of the individual tolerance of the muscular and the cardio-vascular system on the one hand and ergonomically appropriate work organisation on the other. Both factors are verifiable, so that this problem applies equally to women and men and should therefore be answered not by strict regulations but by recommendations (see the shaded areas in table 1).

It goes without saying that the limits shown under "occasionally" must not be exceeded in the range "more frequently".

The word "occasionally" only covers lifting and carrying over a few paces (up to 3 or 4). Carrying over longer distances would have to be included under "more frequently".

The values indicated are not applied to shifting loads by hand. In this case, the limiting factor - which would certainly be below the values indicated here - is the muscle strength, which can be ascertained methodically without difficulty - and the tolerance of the cardio-vascular system. Thus, the admissible load is an individual factor, and it can be adapted by appropriate organisational/technological measures. Shifting of loads should therefore be seen primarily as an ergonomic problem.

The proposed limits for lifting and carrying loads, or lifting strengths, should be considered as binding to protect women against gynaecological damage. For young males, the limits take into account possible damage during their formative years, for which reason maximum loads should also be established. All other values are more or less in the nature of recommendations. Stress damage caused by work itself depends on individual tolerance and the workplace requirements and are thus a problem for industrial medicine and ergonomics. The factory physician will have to ascertain by suitable methods whether a given individual, whether woman or man, can be exposed to higher workplace requirements than those recommended without risking health damage. It goes without saying that the limits presented here, whether binding or recommended, are a compromise. Strictly speaking, there are no exact limits; what would rather be needed are threshold ranges.

- Honduras: Industrial and commercial establishments
  shall take action to limit to 50 kg the weight of any sack or
  package carried by a worker, with a margin
  up to 10 per cent in special cases specified by
  regulations.

- Hungary:

Men over 18 years	Conditions			
	On flat	On 10% slope	On stairs	Other circumstances
Maximum weight				
Distance 90 m	50 kg			
Distance 30 m		50 kg		
Height 3 m			50 kg	

Loads less than 50 kg may be carried over proportionally longer distances. No men are allowed to carry loads to levels higher than 3 m except their own hand tools. On even, non-sagging surfaces, the maximum weights to be carried are:

by hand barrow	100 kg	50 kg	–	loads are to be reduced
by miller cart	200 kg	100 kg	–	according to local condi-
by lever cart	500 kg	200 kg	–	tions and the limits are to
by mine cart	10,000 kg	250 kg	–	be set in the appropriate
by 2-wheeled cart	500 kg	250 kg	–	labour safety code of the
by 3 or 4 wheeled cart	1,000 kg	300 kg	–	particular company

- Poland: The limit for continuous manual lifting and/or carrying on flat surfaces for 25 metres...... 50 kg
On 60° slope (steps) or 45° (ramp) to a maximum height of 4 metres .... 30 kg

2. Adult female workers

(a) Maximum weights for women workers over 18 years old.

In his report on maximum admissible weights for lifting and carrying loads by male, female and young Workers, Professor T. Hettinger studied the frequency distribution curve of limitations on the loads to be carried by women at work for 1970 (figure 1).

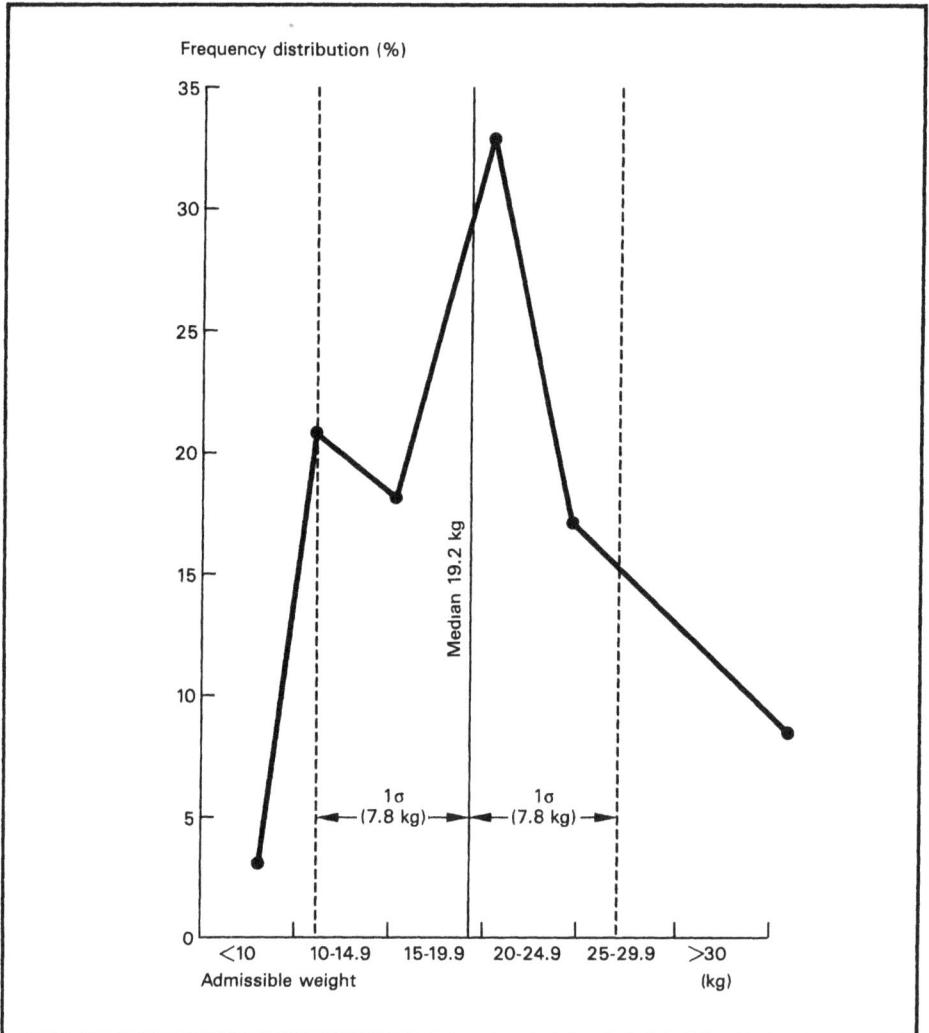

Figure 1: Maximum weight to be carried by women workers over 18 years.

(Compulsory regulations and proposals in various countries.)

Changes have occurred since then, but this does not change the fact, as observed by Hettinger, that many Western and Eastern countries consider such limits as meaningful and that only four of the 53 countries examined had a limit above 35 kg. Taken together, the data gave an average weight limit of about 20 kg.

(b) A greater number of member States adopted limitations of the weight to be manually lifted and carried by women workers than by men.

(c) In some instances, the limitation applies to pregnant women:

- Austria: the federal Maternity Protection Act provides that no expectant mother shall be employed in heavy manual work, including "work in which loads exceeding 5 kg are normally or loads exceeding 10 kg are occasionally lifted by hand without the assistance of machinery, or work in which loads exceeding 8 kg are normally or loads exceeding 15 kg are occasionally moved or transported by hand without the assistance of machinery".

- Luxembourg: during pregnancy and up to three months after delivery, lifting and carrying of loads the weight of which is over 5 kg is forbidden.

(d) In general, limitations apply to non-pregnant women workers. They may fix a maximum absolute load not to be exceeded as in the following case:

- Japan: for women of 18 years and older, in case of:
continuous work .... 20 kg
intermittent work .... 30 kg

- Pakistan (Punjab): for women 17 years of age or older .... 23 kg

- Philippines: maximum weight to be lifted, carried or moved .... 25 kg

(e) In several member States, more elaborate provisions are often in power, such as:

- Bulgaria: the maximum load for manual carrying by women over 16 years of age is:
by hand on an even surface .... 20 kg
by vehicle, not including its weight:
1-wheeled barrow with incline not exceeding 0.02 .... 50 kg
2-wheeled carts on even surface and with incline not
exceeding 0.02 .... 115 kg
2-wheeled carts on uneven rail or paved surface with
maximum incline of 0.01 .... 60 kg
3 or 4-wheeled carts with maximum incline of 0.01 .... 100 kg

When women carry loads by hand, stretcher-type transporters fitted with legs shall be used in so far as the nature of the goods permits; the combined weight of the goods and the transporter shall not exceed 50 kg for the two persons.

- Cameroon: the maximum load to be carried, pulled or pushed by a woman in or outside of the plant is:

by hand .... 25 kg
by carts on rail, including the weight of the vehicle ... 400 kg
by hand barrows, including the weight of the vehicle .... 40 kg
by 3 or 4-wheeled carts, including the weight of the
vehicle .... 60 kg
by 2-wheeled carts (haquets), including the weight of
the vehicle .... 100 kg

- Colombia: for women over 18 years of age,
the maximum load to be lifted may not exceed .... 12.5 kg
to be carried by hand may not exceed .... 15 kg
to be carried on rails may not exceed .... 400 kg
Carrying by means of hand barrow is forbidden to women.

- Côte d'Ivoire: the maximum load to be carried,
pulled or pushed by hand is .... 25 kg
on rails, including the weight of the vehicle .... 600 kg
by wheelbarrow .... 40 kg
by 3 or 4-wheeled carts including the weight of the
vehicle .... 60 kg
by 2-wheeled carts including the weight of the vehicle .. 130 kg

- Czechoslovakia: the maximum load to be lifted
and/or carried by hand by an adult woman is limited at .. 15 kg
For exceptional lifting and/or carrying and if agreed by
the Ministry of Health following a medical examination,
the limit is raised to cycles per shift: 90   16 kg
                                                70   18 kg
                                                44   20 kg
1 cycle means lifting the load up to 1 m and carrying it
at a distance of 10 m. Between the individual cycles,
there must be an interval of 200 per cent of the time of
the duration of the cycle. In such a way, the energy
expenditure due to lifting and carrying during the shift
should not exceed 4.187 kJ (1 Kcal).
Women may not carry loads which exceed:
on single-wheeled cart .... 50 kg
on 2-wheeled cart .... 100 kg
on 4-wheeled cart .... 115 kg
on rails .... 600 kg

- the German Democratic Republic: lifting and carrying regulations for adult women workers take into account, as in the case of adult men workers, both the physical workload and the time spent on lifting during the shift, as shown below:

Frequency of lifting and carrying per shift in % of shift time	Heavy to very heavy work	Moderately heavy work	Light to moderately heavy work
Less than 8 (sporadic)	25 kg	22 kg	19 kg
8-30 (frequent)	15 kg	12 kg	8 kg
More than 30 (constant)	10 kg	7 kg	4 kg

- Federal Republic of Germany: the recommended recognised maximum load is:
for occasional lifting .... 15 kg
for frequent lifting .... 10 kg
(for details see page)

- Hungary: the maximum load to be carried by women over 18 years of age, at a maximum distance of 60 metres, is:
alone .... 20 kg
in pairs .... 40 kg
For shovelling or forking bulk goods, the maximum load, including the weight of the shovel or the fork, is ...... 5 kg
Women may not be employed for mine-car transport.
On slopes not steeper than 2 per cent and on even, non-sinking surfaces, the maximum load to be carried by an adult woman is:
by hand barrow .... 50 kg
by 2-wheeled carts .... 100 kg
by 3 or 4-wheeled carts .... 150 kg
On uneven surfaces or slopes steeper than 2 per cent, the maximum load to be carried by an adult woman is:
by hand barrow .... 20 kg
by 2-wheeled carts .... 50 kg
by 3 or 4-wheeled carts .... 100 kg

- **Malaysia**: women over 18 years shall not carry in excess of <u>25 kg</u> and shall not pull or push loads in excess of the following limits, which include the weight of the vehicle:

trucks on rails	<u>600 kg</u>
wheelbarrows	<u>40 kg</u>
3 or 4-wheeled vehicles	<u>60 kg</u>
small barrows	<u>130 kg</u>

Transport by big two-wheeled barrows and by pedal tricycle carriers is prohibited for women.

- **Mozambique**: women over 21 years of age may carry a weight not exceeding 75 per cent that of a male worker (55 kg, see p. 7). The law states also that the undertakings should avoid as much as possible that adult women are employed in regular carrying of loads. When adult women are employed in regular carrying of loads, the following measures should be adopted: (a) reduction of the time spent on lifting, carrying and putting down of loads, (b) prevention of use of these workers in difficult tasks involved in the manual transport of loads.

- **Poland**: in case of permanent lifting and/or carrying, the maximum load for an adult woman is:

on flat	<u>20 kg</u>
going up on a slope	<u>15 kg</u>

In case of occasional lifting and/or carrying the maximum load is:

on flat	<u>30 kg</u>
going up on a slope	<u>25 kg</u>

- **Thailand**: the maximum load to be lifted, carried on the shoulder, carried on the head, pulled or pushed by a woman on level ground is

a woman on level ground is	<u>30 kg</u>
ascending stairs or elevated places	<u>25 kg</u>
pulling or pushing on wheels with rails	<u>600 kg</u>
pulling on wheels without rails	<u>300 kg</u>

- **USSR**: for women manually lifting and/or carrying loads, the maximum weight varies as follows: lifting and carrying loads alternating with other

operations	<u>15 kg</u>
lifting loads over the height of 1.5 metres	<u>10 kg</u>
lifting and carrying loads continously	<u>10 kg</u>

The total amount of loads lifted and carried during the shift shall not exceed 7,000 kg.
The weight of the load lifted and/or carried includes the weight of the container and package.
When carrying loads on carts or in standard containers, the effort should not exceed 15 kg.

3. Young working persons and children

The frequency distribution curve of limitations on the loads to be carried by young persons at work for 1970 as calculated by Professor T. Hettinger is shown below (figure 2).

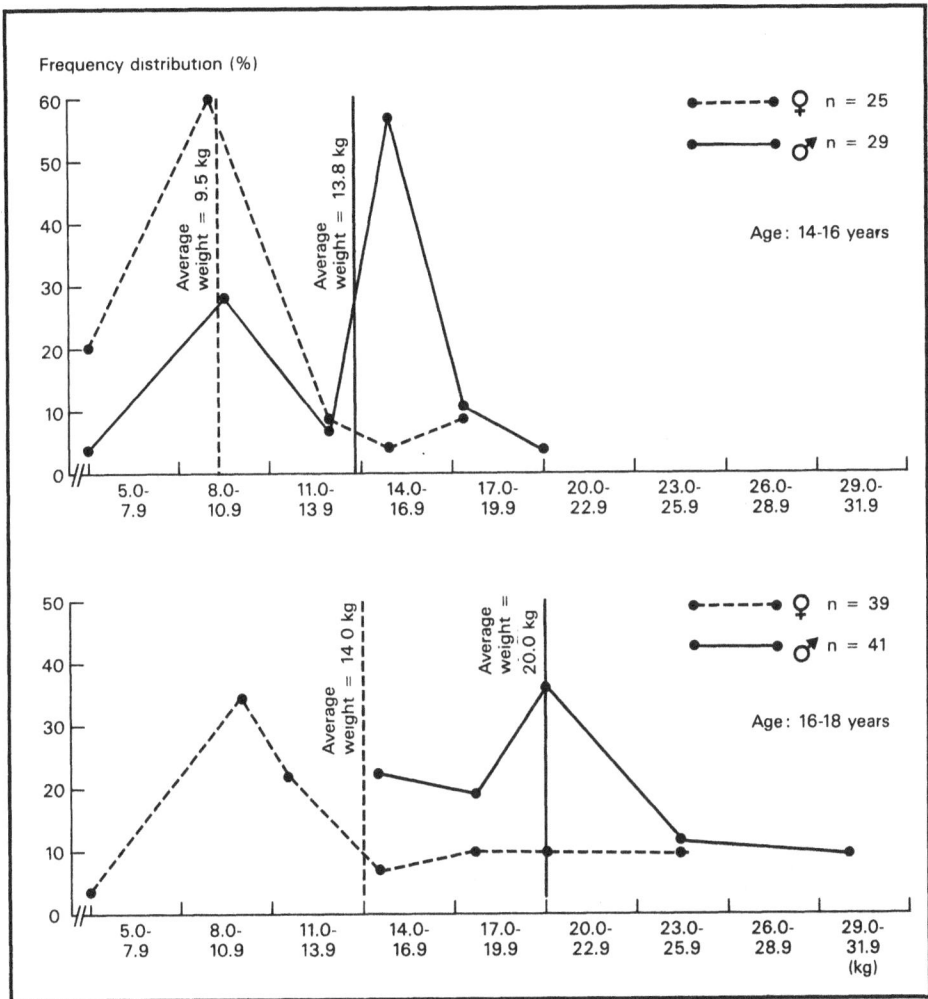

Figure 2: Maximum weights for young people in various countries. (n = number of countries)

The two groups (young people between 14 and between 16 and 18 years of age) showed widely differing opinions about the admissible limits, although most of the data - more than 85 per cent in each case - were in a range of about 4.5 kg above and below the median value.

The following limits are referred to:

- Bolivia: The following weights should not be
  exceeded in manual transport of loads:
  by hand, boys under 16 years of age ....   10 kg
  "   "   girls   "   "   "   "   "   ....   5 kg
  "   "   women 16-20 years of age ....   10 kg
  on rail, boys under 16 years of age ....   300 kg
  "   "   women 16-20 years of age ....   300 kg
  "   "   girls under 16 years of age ....   150 kg
  by handcarts, boys 14-16 years of age ....   40 kg
  by 3 or 4-wheeled carts, women 18-20 years of age .   50 kg
  "   "   "   "   girls under 18 years of age .   35 kg
  "   "   "   "   boys under 16 years of age ..   35 kg

- Bulgaria: The limits set down for manual handling
  of loads by women (see p. 15) also apply to girls
  16-18 years of age.

- Colombia: For intermittent lifting by young working
  persons between 16 and 18 the maximum load is......   15 kg
  For continuous lifting it is 25 per cent less,
  that is... 11.250 kg
  Carrying by hand is limited as follows:
  16-18 years ....   20 kg
  boys up to 16 years ....   15 kg
  girls "   "   "   ....   8 kg
  For carrying by carts on rail the maximum load is:
  16-18 years of age ....   500 kg
  up to 16 years, boys ....   300 kg
  "   "   "   "   girls ....   200 kg
  For carrying by hand barrow, the maximum load is:
  18 years ....   40 kg
  16-18 years ....   20 kg

- Côte d'Ivoire: The maximum weights to be manually
  carried, pulled or pushed by adolescent workers
  both inside and outside the usual workplace are
  determined as follows:
  by hand, young workers 16-18 years of age..........   20 kg
  "   "   boys 14-16 years of age ....   15 kg
  by carts on rail, including the weight of
  the vehicle:
  boys 14-17 years of age ....   500 kg
  girls 16-17 "   "   "   ....   300 kg
  girls under "   "   "   ....   150 kg
  by hand barrow, including the weight of the
  vehicle:
  boys 14-17 years of age ....   40 kg

by 3 or 4-wheeled carts, including the
weight of the vehicle:
boys 14-17 years of age .... 60 kg
girls 17-18 "   "   "   .... 60 kg
girls under 16 years of age .... 35 kg
by two-wheeled carts, including the weight
of the vehicle:
boys 14-17 years of age .... 130 kg
by tricycles, including the weight of the vehicle
boys 16-17 years of age .... 75 kg
 "   14-15   "   "   "   .... 50 kg

- Czechoslovakia: For occasional transport of loads,
  the following maximum limits are specified by law:
  by hand, boys and girls up to 16 years of age ..... 10 kg
  on a carrier in pairs, boys and girls up to 16 years
  of age .... 20 kg
  on a 4-wheeled cart, boys and girls up to 16 years
  of age .... 50 kg
  by hand, young workers 16-18 years of age.......... 20 kg
  on a carrier in pairs, young workers 16-18 years
  of age .... 50 kg
  on a 4-wheeled cart, young workers 16-18 years of age 100 kg
  by hand, girls, 16-18 years of age .... 13 kg
  on a carrier in pairs, girls, 16-18 years of age ... 30 kg
   " " 4-wheeled cart   "   "   "   " "..... 65 kg
  For permanent or repeated lifting, carrying and
  unloading, governmental guide-lines authorise
  ministries to fix the maximum permissible weight
  depending upon the number of operations and other
  conditions.

- Ecuador:
  age 16-18, male .... 23 kg
   " 18-21, female .... 12 kg
  below 16 years, male .... 16 kg
   "   18   "   female .... 9 kg

- Egypt: The maximum load to be manually carried is:
  boys between 12 and 15 years of age .... 10 kg
  girls   "   "   "   "   "   " .... 7 kg
  The maximum load to be pushed or pulled on rails is:
  boys between 12 and 15 years of age .... 30 kg
  girls   "   "   "   "   "   " .... 15 kg

- Finland:
  age 15-17, male .... 20 kg
   "   "   female .... 15 kg

- Gabon: The maximum weights in manual transport of
  loads by adolescent workers have been fixed as
  follows:
  by hand, boys 17-18 years of age ....   20 kg
  "   "   16-17   "   "   "   ....   15 kg
  "   " girls 17-18   "   "   "   ....   10 kg
  by wheelbarrow (weight of the vehicle included):
  boys 17-18 years of age ....   45 kg
  " 16-17   "   "   "   ....   35 kg
  girls 17-18   "   "   "   ....   35 kg
  by 3 or 4-wheeled carts (weight of the cart included):
  boys 17-18 years of age ....   50 kg
  " 16-17   "   "   "   ....   45 kg
  girls 17-18   "   "   "   ....   45 kg
  on rail (weight of the truck included):
  boys 17-18 years of age .... 500 kg
  " 16-17   "   "   "   .... 400 kg
  girls over 16 years of age .... 300 kg
  "   aged 16 .... 150 kg

- Federal Republic of Germany:
  age 15-18, male occasionally ....   35 kg
  "   "   "   frequently ....   20 kg
  age 15-18, female occasionally ....   15 kg
  "   "   "   frequently ....   10 kg
  (for details see page 8)

- Greece: The following maximum limits of loads were
  adopted with regard to young workers and children:
  lifting and carrying by hands, young workers 16-18   10 kg
  "   "   "   "   "   children up to 16 ...   5 kg
  pushing on rail, young workers 16-18 years of age .. 300 kg
  carrying by means of a wheelbarrow, 16-18 years, males 50 kg

- Hungary: The following limits are specified for the
  occasional lifting and carrying of loads by
  adolescents:
  Young workers, 16-18, on flat for a maximum distance
  of 60 m:
  alone ....   20 kg
  in pairs ....   40 kg
  boys 14-16 years of age on flat, alone ....   15 kg
  "   "   "   "   "   "   "   in pairs ....   30 kg
  girls 16-18   "   "   "   "   "   alone ....   15 kg
  "   "   "   "   "   "   "   "   in pairs ......   30 kg
  boys 14-16 and girls 16-18 on ramp with 1% slope ...   10 kg
  "   "   "   "   "   "   "   "   2%   "   ...   5 kg
  Boys between 14-16 and girls between 16-18 years of
  age may carry on slopes not steeper than 2 per cent
  the following maximum weight:

```
by 2-wheeled carts ..................................    50 kg
by 3 or 4-wheeled carts ............................    70 kg
On slopes steeper than 2 per cent or on uneven
surfaces:
by 2-wheeled carts ..................................    30 kg
by 3 or 4-wheeled carts ............................    50 kg
```

- Israel: The following limits were adopted with
  regard to manual transport of loads by young persons
  and children:

```
16-18 years, male, if the shift does not exceed
                     2 hours a day ...................    20 kg
  "      "     "    if the shift exceeds 2 h/day .....    16 kg
under 16 years, male, if the shift does not exceed
                     2 hours a day .................   12.5 kg
  "    "     "     "    if the shift exceeds 2 h/day...    10 kg
girls 16-18 years, shift not to exceed 2 h/day ......    10 kg
  "   under 16 years, shift not to exceed 2 h/day ...     8 kg
For transport by means of wheelbarrow, including the
weight of the vehicle, boys 16-18 years .............    50 kg
```

- Japan: The limits of load to be lifted and/or
  carried by adolescent workers differ whether the
  work is intermittent or continuous:

```
intermittent, age 16-18 years, male .................    30 kg
     "         "    "    "    female ...............    25 kg
     "        under 16 years, male ..................    15 kg
     "         "    "    "    female ...............    12 kg
continuous, age 16-18 years, male ...................    20 kg
     "        "    "    "    female ...............    15 kg
     "       under 16 years, male ...................    10 kg
     "        "    "    "    female ...............     8 kg
```

- Malaysia: Young persons and children shall not
  carry loads exceeding the following weights:

```
16-18 years of age, male ............................    20 kg
  "    "    "    "    female .........................    10 kg
14-16 years of age, male ............................    15 kg
  "    "    "    "    female .........................     8 kg
They shall also not pull or push loads exceeding the
following weights, which include the weight of the
vehicle:
by trucks on rail, 14-17 years of age, male ........   500 kg
  "     "     "     "    16-17  "   "   "   female .......   300 kg
  "     "     "     "    under 16 years of age, female ....   150 kg
by wheelbarrows, 14-17 years of age, male ...........    40 kg
by 3 or 4-wheeled carts, 14-17 years of age, male ...    60 kg
  "   "    "     "    over 16  "   "   "   female ..    60 kg
  "   "    "     "    under 16 years of age, female     35 kg
by small barrows, 14-17 years of age, male ..........   130 kg
by tricycle carriers, 16-17 years of age, male ......    75 kg
  "     "     "    14-17  "   "   "   "   "   ......    50 kg
```

- <u>Mexico</u>: For manual transport of loads on the back or by hand, the maximum weight is specified as follows:

boys under 16 years of age .... <u>20 kg</u>
girls "   "   "   "   " .... <u>10 kg</u>

Adolescent workers may not push or pull loads requiring a muscular effort exceeding that necessary to move horizontally the following weights on rail, including the weight of the vehicle:

boys 14-16 years of age .... <u>400 kg</u>
girls "   "   "   " .... <u>250 kg</u>
boys under 14 years of age .... <u>200 kg</u>
girls "   "   "   "   " .... <u>150 kg</u>
by wheelbarrow, boys 14-16 years of age .... <u>40 kg</u>
by 3 or 4-wheeled vehicles, boys 14-16 years of age. <u>50 kg</u>
"   "   "   "   girls "   "   "   " ... <u>40 kg</u>
"   "   "   "   boys under 14 years of age <u>30 kg</u>
"   "   "   "   girls "   "   "   "   " <u>20 kg</u>
by pedal tricycle carriers, boys 14-16 years of age <u>50 kg</u>

Children under the age of 16 shall not be employed for more than four hours during the working day in continuously carrying the weights mentioned above.

- <u>Pakistan</u> (Punjab): Lifting and carrying loads is forbidden to adolescents in excess of:

age 15-17, male .... <u>23 kg</u>
"   "   female .... <u>18 kg</u>
" 15 and below .... <u>16 kg</u>

- <u>Poland</u>: Adolescents and young workers shall not carry loads exceeding the following limits:
young workers 16-18 years on slope, with a maximum gradient of 30° and a maximum height of 5 m .... <u>8 kg</u>
boys under 16 years of age, on flat .... <u>16 kg</u>
"   "   "   "   "   "   " slope, steps or ramp <u>5 kg</u>
girls under 16 years of age, on flat .... <u>10 kg</u>
"   "   "   "   "   "   " slope, steps or ramp <u>3 kg</u>

IV. No limiting provisions

As it was explained in the introduction, the present document cannot pretend to be exhaustive because the information available was not complete, and often not official. However, it may be of interest to know in which member States legislation on maximum weights does not exist for the time being.

To date, the following countries can be included in such a group:

Chile (only general provisions)
Cuba
Ethiopia
Indonesia (only general provisions)
Kenya
Papua New Guinea
Portugal
Sierra Leone
Singapore
Solomon Islands
Swaziland
United Republic of Tanzania

V. Practice of load lifting and carrying
at the workplace

1. **In several member States**, limitations of load lifting
and carrying adopted early in this century are at present
generally obsolete or replaced by good working practice. The
following countries or areas may be included in this group (in
brackets, the date of the enactment of the relevant
legislation): France (1909), the USSR (1921), Switzerland
(1923), Chile (1923), the United Kingdom (1926), Malta (1926),
the State of South Australia (Australia) (1926), Belgium (1926),
the State of New South Wales (Australia) (1927), Portugal (1927),
Hong Kong (1932), Cyprus (1932), Italy (1934), Mexico (1934) and
several states of the United States.

2. **In industrialised countries**, the practice of manual
load transport has greatly diminished as a result of
mechanisation, and has greatly improved as a result of new
physiological, biomechanical and ergonomic knowledge with special
regard to static and dynamic stress imposed on the spine in
relation to posture. The causes of back injuries and other
health problems associated with physical activity are better
known and the value of training and correct kinetic techniques in
lifting and carrying have been recognised. The concept of
maximum load has become more complex because in addition to the
weight lifted and carried, other factors should be considered,
such as the distance it is carried, the slope involved, the rate
of energy expenditure, the percentage of worktime spent on manual
transport and the total workload of the workshift. Other
factors such as the size of the load, the ease of grip, the
symmetry of effort and the ambient temperature also play a role
in the evaluation of the physical effort.

By January 1987, 22 member States had ratifed the Maximum
Weight Convention, 1967 (No. 127).

A number of member States have adopted and widely publicised
codes of practice and guide-lines on manual transport of loads
(such as Australia, Austria, the United Kingdom). It is not the
purpose of this document to review or summarise these
contributions. However, in consideration of its semi-official
character the conclusions of Code of Practice 415 on Manual
Handling, issued by the Australian Government Publishing Service
(tables B1 and B2), are reported below.

Table 2: Acceptable weight of lift (kg). All lifts are up to 760 mm vertical lifts, once every 8 hours (ref: S.H. Snook – The design of manual handling tasks ergonomics, 1978, Vol. 21, No. 12, 963–985)

	Centre of gravity of object from body	Acceptable weight	Floor level to knuckle height	Knuckle height to shoulder height	Shoulder height to arm reach
Males	380 mm	Optimum	23	19	18
		Maximum	29	24	23
	250 mm	Optimum	26	19	18
		Maximum	34	24	23
	180 mm	Optimum	29	20	19
		Maximum	37	26	24
Females	380 mm	Optimum	17	13	12
		Maximum	20	15	14
	250 mm	Optimum	20	13	12
		Maximum	24	15	14
	180 mm	Optimum	22	14	13
		Maximum	26	17	15

Table 3: Weights for frequent lifting (expressed as a percentage of the acceptable weight)

Frequency of lifting	Percentage of acceptable weight
Once in 30 minutes	95
Once in 5 minutes	85
Once in 1–2 minutes	66
Once in 10–15 seconds	50
Once in 5 seconds	33

Notes

1. The optimum loads can be handled by 90 per cent of workers and should be used to plan all routine and repetitive manual handling tasks.

2. The maximum permissible weights are those that 75 per cent of healthy workers aged between 18 and 60 years can be expected to lift using any freely chosen lifting technique. It is expected that loads at these levels would be handled infrequently and only with the authorisation of a responsible officer taking care in the selection of persons of appropriate physique, fitness and experience.

3. Workers aged 60 years and over should not be required to lift more than the optimum loads.

4. The figures given in the tables should be reduced by 25 per cent for workers between the ages of 16 and 18 years and 50 per cent for workers less than 16 years of age.

5. No woman who is pregnant should be assigned to the manual handling of loads greater than those described as optimum and as qualified by the frequency of lifting.

Another approach to the regulation of manual transport of loads is found in the Proposals for Health and Safety (Manual Handling of Loads) Regulations and Guide-lines of the United Kingdom Health and Safety Commission. The action levels relevant to handling tasks involving the regular manoeuvring of fairly compact loads by an individual in a good environment are indicated in table 1 (Action levels for employers).

Table 4: Action levels for employers

Levels	Actions
Below 16 kg	No special action required, provided those relatively few individuals likely to face serious risks when handling weights of this order have been identified.
From 16 kg to 34 kg	Administrative procedures required to identify those individuals unable to handle such weights regularly without unacceptable risk, unless mechanical assistance is provided.
Above 34 kg to 55 kg	Unless the regular handling of weights of to 55 kg these magnitudes is limited to effectively supervised, selected and trained individuals, mechanical handling systems should be employed.
Above 55 kg	Mechanical handling systems should always be considered at this level. Where not reasonably practicable, selective recruitment and special training is essential, since even with effective supervision, very few people can regularly handle weights of this order with safety.

Collective agreements with or without the participation of governmental authorities should be not overlooked in this area because they have a great practical value and reflect real working conditions.

Thus, in <u>Finland</u>, negotiations under the aegis of the Division of Forestry and Agriculture of the National Board of Labour Protection led to the reduction of the weight of sacks of fertilisers and seed grain – and later on of cement – from 50 to 40 kg, as soon as it was technically possible on the packing lines. Bigger amounts of fertilisers are packed in 600 kg sacks and are mechanically handled.

In <u>Japan</u>, 55 kg as the maximum weight of loads being able to be handled by a male worker has been established as a guiding principle, although it does not have the actual effect of legal prohibition. When a weight more than 55 kg is handled, two workers are engaged in the transport and each worker shares the weight equally.

In New Zealand, the agreements and awards made under the Industrial Relations Act, 1973, usually repeat the general legislative requirement (see p. 4). In a few cases only, maximum permissible weights for women and young workers are prescribed, from a maximum of 13 kg for women workers to a maximum of 32 kg (unassisted) for young workers.

3. In developing countries, the practice of manual transport of loads is not so advanced from the point of view of health protection as in the industrialised countries. Information available to the ILO is as follows:

In Bangladesh, the maximum load carried on the head or neck is 90 kg.

In China, the maximum load carried by an adult male worker is 80 kg. However, over 50 kg the range is limited to 70 m.

In Colombia in the construction industry, according to a resolution of 1979, the maximum weight should not exceed 50 kg for adult workers in occasional lifting and should be 25 per cent less in continuous lifting.

In India at docks the maximum weight is still 100 kg. Furthermore, in the non-organised sector loads as heavy as 115-135 kg are manually carried by an individual worker.

In the Republic of Korea, the rice packages do not exceed 40 kg.

In Malaysia on oil-palm estates, the average size fruit bunch weighs 20 kg; the worker carries two baskets containing two to three bunches each, the total load being between 80 and 120 kg. In a working day he can manually load 500 to 600 bunches weighing altogether some 10-12 tons. On rubber estates, the manual carrying of latex is made by means of two buckets hooked on either end of a stick, the total load being up to 60 kg carried over a distance of 2-3 km. In chemical spraying, the conventional sacksprayer carried on the back of the worker weighs 20 kg with a full load.

In Mexico, the recommendations of the General Directorate of Occupational Health and Safety are not to exceed 56 kg at a maximum height of 1.50 m in lifting 1-2 times per hour, 40 kg in repeated lifting 1-6 times every 15 minutes; 80 kg as the maximum weight, always at a maximum height of 1.50 m. For women the corresponding values but at a maximum height of 1.30 m are respectively 27 kg, 19 kg and 38 kg. Young persons 16-18 years of age should not exceed 19 kg (male) or 12 kg (female), boys

14-16 years of age should not exceed <u>16 kg</u> and girls <u>10 kg</u>. Workers more than 50 years of age should not carry or lift more than <u>16 kg</u> (men) or <u>10 kg</u> (women).

In <u>Sierra Leone</u> it is common practice for workers to carry 50 kg bags of rice or palm kernels on their heads.

VI. References and sources

AUSTRALIA. Occupational Safety and Health in Commonwealth Government Employment – Code of practice, 415 Manual Handling, Australian Government Publishing Service, Canberra 1982.

[New South Wales. Factories and Shops (Amendment) Act, 1927, from Report I, PCPTW/1966/I/ILO.]

[South Australia. Industrial Code of 1926, from Report I, PTCPW/1966/I/ILO.]

AUSTRIA. 218 Verordnung des Bundesministers für soziale Verwaltung vom 11 März 1983 über allgemeine Vorschriften zum Schutz des Lebens, der Gesundheit und der Sittlichkeit der Arbeitnehmer (Allgemeine Arbeitnehmerschutzverordnung – AAV) Bundesgesetzblatt für die Republik Osterreich. Jahrg. 1983, 7 April 1983,

527 Verordnung der Bundesminister für soziale Verwaltung und für Handel, Gewerbe und Industrie vom 2 Oktober 1981 über die Beschäftigungsverbote und –beschränkungen für Jugendliche. Bundesgesetzblatt für die Republik Osterreich. Jahrg. 1981, 3 Dezember 1981,

(Federal) Maternity Protection Act, 1957, from Report I, PTCPW/1966/I/ILO.

BANGLADESH. Information collected at the International Symposium on Ergonomics in Developing Countries, Jakarta, 18–21 November 1985.

BELGIUM. Royal Order of 3 May 1926 relating to the employment of women and children, from Report I, PTCPW/1966/I/ILO.

BOLIVIA. Regulations for the administration of the Presidential Decree of 21 September 1929 respecting the protection of women and children in industry, from Report I, PTCPW/1966/I/ILO.

BRAZIL. Information presented at the Symposium on Ergonomics in Developing Countries, Jakarta, 18–21 November 1985.

BULGARIA. Employment of Women Ordinance No. 53 of 3 July 1959, from Report I, PTCPW/1966/I/ILO.

CAMEROON. Arrêté Ministériel 16 du 27 mai 1969 du Ministère du Travail et de la Prévoyance Sociale (MPTS).

CANADA. Canada Materials Handling Regulations under Canada Labour Code. CRC 1978 c. 1004 as amended by SOR/80-302 gazetted 14 May 1980; SOR/80-675 gazetted 10 September 1980; SOR/84-534 gazetted 25 July 1984.

[Quebec. Safety Code for the Construction Industry under an Act respecting occupational health and safety. SQ 1979 c. 63 after consolidation RSQ c.S.21; as amended by Notice of Replacement gazetted 8 February 1984.]

CHILE. Act No. 3915 of 9 February 1923, from Report I, PTCPW/1966/I/ILO.

Information provided by the Departamento de Salud Ocupacional y Contaminación Ambiental. Ministerio de Salud, 16 January 1987.

CHINA. Information provided by Quian Heng, Department of Labour Protection, Beijing College of Economics, at the Symposium on Ergonomics in Developing Countries, Jakarta 18-21 November 1985.

COLOMBIA. Estatuto de Seguridad Industrial, Resolución 02400 of 22 May 1979, Ministerio del Trabajo y Seguridad Social, articles 392, 698, 107, 108, ibidem, De la Ergonomia en la Construcción, articles 82, 83.

COTE D'IVOIRE. Décret no. 67-265 du 2 juin 1967, Livre III, Code du Travail, articles 3D 305 and 3D 333.

CUBA. Reglamento General de la Ley de Protección e Higiene del Trabajo, Cap. XII, article 123.

CYPRUS. Employment of Children and Young Persons Law, 1932, from Report I, PTCPW/1966/I/ILO.

CZECHOSLOVAKIA. Hygiene Regulations 36/1976. "Guide-lines of hygiene demands on stationary machines and technical equipment."

Government Decree 32/1967. "Principles for the list of operations and workplaces forbidden for women, pregnant women and mothers until the end of the ninth month after delivery and adolescents."

ECUADOR. Reglamento de Seguridad e Higiene del Trabajo. Resolución No. 172, IESS, 1975.

EGYPT. Ministerial Decree No. 12 of 1982 in application of article 145 of the Labour Law set forth under Law No. 137 of 1981. Article 1, paragraph 2.

Ministerial Decree No. 13 of 1982 in application of article 145 of the Labour Law set forth under Law No. 137 of 1981. Article 1, paragraphs 1 and 22.

Ministerial Decree No. 22 of 1982 in application of article 153 of the Labour Law set forth under Law No. 137 of 1981. Article 1, paragraphs 3, 19 and 20.

ETHIOPIA. Information received from the Labour Inspection Division, 28 June 1985.

FINLAND. Information provided by the National Board of Labour Protection, letter of 9 February 1987.

FRANCE. Decree of 28 December 1909, amended on 26 October 1912.

Circulaire du 2 Mai 1985 publiée au Bulletin Officiel du Ministère du travail de l'emploi et de la formation professionnelle, No. TR 85/23-24 du 2 juilliet 1985.

GABON. Decree No. 275 of 5 December 1962, from Report I, PTCPW/1966/I/ILO.

GERMAN DEMOCRATIC REPUBLIC. Handbuch für den Gesundheits- und Arbeitsschutz. Tribüne, Berlin 1976, p. 467.

FEDERAL REPUBLIC OF GERMANY. Recommendation by the Federal Ministry of Labour and Social Affairs of 1 October 1981; T. Hettinger, Heben und Tragen von Lasten – Der Bundesminister für Arbeit und Sozialordnung – 1981.

GREECE. Decree 3926/65 presented at the Symposium on Ergonomics in Developing Countries, Jakarta, 18-21 November 1985.

Law concerning women and children, referred to as above.

GUATEMALA. General Regulations on Occupational Hygiene and Safety, 28 December 1957, from Report I, PTCPW/1966/I/ILO.

HETTINGER, T. See under Federal Republic of Germany.

HONDURAS. Decree No. 189 of 1 June 1959 (Labour Code).

HONG KONG. Factories and Workshops Ordinance No. 27 of 1932, from Report I, PTCPW/1966/I/ILO.

HUNGARY. Decree 2/1972 MK6, KPM of the Ministry of Transport and Communication, letter of 26 February 1987 from the National Research Institute of Occupational Safety.

INDIA. Information presented at the Symposium on Ergonomics in Developing Countries, Jakarta 18-21 November 1985.

INDONESIA. Act No. 1 of 1951 to bring Labour Act No. 12 of 1948 of the Republic of Indonesia into operation throughout Indonesia, State Gazette No. 2 of 1951.

IRELAND. Factory Act of 1955, from Report I, PTCPW/1966/I/ILO.

ISRAEL. Order of 15 January 1954, from Report I, PTCPW/1966/I/ILO.

ITALY. Act 653 of 26 April 1934, from Report I, PTCPW/1966/I/ILO.

JAPAN. Ordinance to regulate the employment of women and children No. 13 of 19 June 1954.

Guide-lines for Lumbago Prevention Measures in Heavy Weights Handling Works provided by the Institute of Industrial Health, Ministry of Labour, Japan.

KENYA. Information provided by the Ministry of Labour, 26 June 1985.

REPUBLIC OF KOREA. Information presented at the Symposium on Ergonomics in Developing Countries, Jakarta, 18-21 November 1985.

LUXEMBOURG. Loi du 3 juillet 1975 concernant: (1) la protection de la maternité de la femme au travail, (2) la modification de l'article 13 du code des assurances sociales modifié par la loi du 2 mai 1974.

Loi du 28 octobre 1969 concernant la protection des enfants et des jeunes travailleurs.

MALAYSIA. Decree No. 62/152 quoted in Report I, PTCPW/1966/I/ILO.

Information presented at the Symposium on Ergonomics in Developing Countries, Jakarta, 18-21 November 1985.

MALTA. Factories' Regulation Act, 1926, mentioned in Report I, PTCPW/1966/I/ILO.

MEXICO. Regulations respecting employment of children in dangerous and unhealthy occupations, 31 July 1934.

Posturas y movimientos en el trabajo. Confort en la industria. Dirección General de Medicina y Seguridad del Trabajo, Departamento de Ergonomia.

MOZAMBIQUE. Dispatch dated 12 December 1972. Residencia do Governo Geral de Moçambique. O Secretário provincial do Trabalho, Previdência e Acçáo Social.

Decree No. 36/76 dated 26 October 1976, Cap. XIV, Seç 1, Trabalho de mulheres.

NETHERLANDS. Labour Youth Decree, sections 7 and 8.
1920 Labour Decree according to a letter from the Division of the International Social Affairs dated 10 July 1985.

NEW ZEALAND. Factories and Commercial Premises Act, 1981, No. 25. Section 26: Carrying of heavy loads.

PAKISTAN. Factories Act, 1934, article 33F.

[Punjab. Factories Rules 1978, article 78.]

PHILIPPINES. Occupational Safety and Health Standards promulgated by the Minister of Labor and Employment pursuant to the Labor Code of the Philippines P.D.422, Rule 1410.

POLAND. Order of the Minister of Labour and Social Welfare of 1 April 1953 on occupational health and safety during manual handling materials. Dziennik Ustaw No. 22 poz. 89.

Order of the Cabinet of 28 February 1957 on works forbidden to women. Dziennik Ustaw No. 12 poz. 96.

Order of the Cabinet of 26 September 1958 on works forbidden to adolescents. Dziennik Ustaw No. 64 poz. 312.

PORTUGAL. Decree 17/84 of 4 April 1984 to ratify Convention No. 127.

Decree 14/498 of 29 October 1927, from Report I, PCTPW/1966/I/ILO.

SIERRA LEONE. Information presented at the Symposium on Ergonomics in Developing Countries, Jakarta, 18-21 November 1985.

SINGAPORE. Information provided by the Department of Industrial Health, Ministry of Labour.

<u>SOLOMON ISLANDS</u>. Safety at Work Act, 1982.

<u>SRI LANKA</u>. Factories Ordinance, section 58, as mentioned in letter of 30 July 1985 from the Ministry of Labour.

<u>SWAZILAND</u>. Information provided by the Department of Labour.

<u>SWEDEN</u>. Ordinance AFS 1983/6 of the National Board of Occupational Safety and Health.

<u>SWITZERLAND</u>. Factory Act, Amendment of 17 September 1923, from Report I, PTCPW/1966/I/ILO.

<u>UNITED REPUBLIC OF TANZANIA</u>. Information provided by the Ministry of Labour.

<u>THAILAND</u>. Notification of the Ministry of Interior re Labour Protection clause 14.

<u>USSR</u>. Decision of the State Committee of the USSR for Labour and Social Affairs and the Presidium of the All-Union Central Council of Trade Unions No. 22, P-I, of 27 January 1982, on Weight Limit Standards for Women Lifting and Carrying Loads Manually.

<u>UNITED KINGDOM</u>. Woollen and Worsted Textiles (Lifting and Heavy Weights) Regulations, 1926, referred to in Report I, PTCPW/1966/I/ILO.

Proposals for Health and Safety (Manual Handling of Loads) Regulations and Guidance, HMSO, London, 1982.

"Manual handling and lifting: An information and literature review with special reference to the back", J.D.G. Troup and F.C. Edwards, HMSO, London, 1985.

VII. Index by country or area